THE NON-ZERO UNIVERSE

A NEW LOOK AT THE LOGIC
OF A COSMOLOGICAL EVERMORE

WITH A BRIEF OVERVIEW
OF ITS LINKS IN HISTORY

JAMES D ROBINSON

EOCRANTIS

First published in 2020
by : Eocrantis Publishing
10 Bewick Crescent
Newton Aycliffe
Co Durham
DL5 5LQ
England

ISBN : 978-0-9571378-7-5

Cover design : James D Robinson
Background cover photo : via Pixabay Free Photos / 9091

CONTENTS

James D Robinson was born in Darlington, Co Durham, England, in 1949, Apart from writing, he has a broad range of other interests, including: photography, drawing, graphic design - and cosmology (hence this book.)

His photo-work is represented in the UK's permanent photo collections of both: The National Portrait Gallery, and The Picture Library of the Royal Society, and his writing and photographs have been published in: Popular Astronomy magazine, The Best of British magazine, The Journal of the Royal Photographic Society, Amateur Photographer magazine, What Digital Camera magazine, Skermish magazine, and: North East Life magazine.

He has also self published a number of books, including: The Authenticated Meteoric Falls of the British Isles, The Pseudo-Meteoric Events of the British Isles, and The Day the Bridges Fell, (an historical account of a devastating flood which brought death and destruction to the North East of England in 1777,) and additionally, he is the creator of a time capsule, now stored in the archive department of the UK's Durham County Records Office, where it will remain unopened until the 21st of March, 2317 - making it, at the time of this writing, the 5th longest 'sealed' time capsule on record.

"Not only is the universe stranger than we think,
it is stranger than we can think."

(Werner Heisenberg : 'Across the Frontiers')

PREFACE

The mysteries of the universe have intrigued and perplexed humanity since the earliest of times. Our knowledge of the natural order of things - both on and off planet Earth - has increased substantially, and yet some ancient questions still arouse much debate, the biggest of which remains: Did the universe have a beginning or has it always existed?

This book, short though it is, will attempt to offer a possible answer, not by turning to the spiritual dogmas of religion or to the more outlandish complexities of modern science, but by the use of simple logic and common sense intuition, for if we divest ourselves of both, we fall into the realms of magic and fantasy and remove ourselves from whatever amounts to the true reality.

WHEN THE GODS RULED & BEYOND

To the brightest minds of the ancient world; often scholars who made interpretations tied to religious beliefs, the universe was first considered to be an endless ocean[1] before the earth came into being via divine creation.

Later, after the dawn of 'living things,' the Hebrews pictured the earth, seated in the centre of the ocean, as a thick circular[2] disc with lofty mountains around its perimeter, whilst down in the bowels of the earth where darkness reigned, a huge chamber was envisioned representing the 'underworld,' into which the spirits of the living passed after death.

Deeper down still, and extending outwards into the waters of the abyss, pillars of rock were believed to support the weight of the world above, whilst the sky was held to be the vault of an enormous dome which rose upwards from around the very edge of the world, its purpose being to keep apart the earth from the endless ocean - which existed even above the dome - except when allowed to fall as rain via gates or windows in the dome, - as and when the gods decreed.

The sky-dome itself (according to various records) is usually described as being made of some type of metal or stone,[3] though it has also been referred to in more poetic ways, such as being like a tent stretched across the firmament,[4] (which is one way the Egyptians saw it,) though in their configuration, the tent was held aloft by the tops of mountains which marked the four corners of the known world. The remaining space, between the sky-dome (or tent) and the land below, contained the breathable air which thus permitted life, though the ancient Babylonians took a different view overall.

Although their earth was still considered 'the central place' in the universe, they saw it as being of two halves, an upper half associated with light and life, and a lower half associated with darkness and death. Each half had some semblance to a seven staged pyramid,[5] with the peak of the upper half being described as "like a mountain penetrating the clouds." Their earth was also thought to be 'four cornered' [5] although to gain entry to the "abode of the dead," one had to cross a body of water which separated the two halves. (See the illustrations on the following pages.)

Whilst the above concepts, seem strange to our minds, they do appear to have been the general design for most early civilizations, reaching back to the ancient Mesopotamians who had developed hilltop cities from around 9,000 BC.[6]

Even so, in the 6th cent BC, Anaximenes of Miletus, suggested the earth was a cylinder with a width three times the dimension of its height.[7]

This structure, he surmised, was encircled by three rings, one inside the other with the diameter of the rings being: 9, 18 and 27 times that of the earth. The outermost ring carried the fixed stars, whilst the other two accommodated the motions of the sun and moon. Later, when other teachings from the Greek scholars became more highly influential, the celestial bodies, which previously had been embedded into the ground-facing

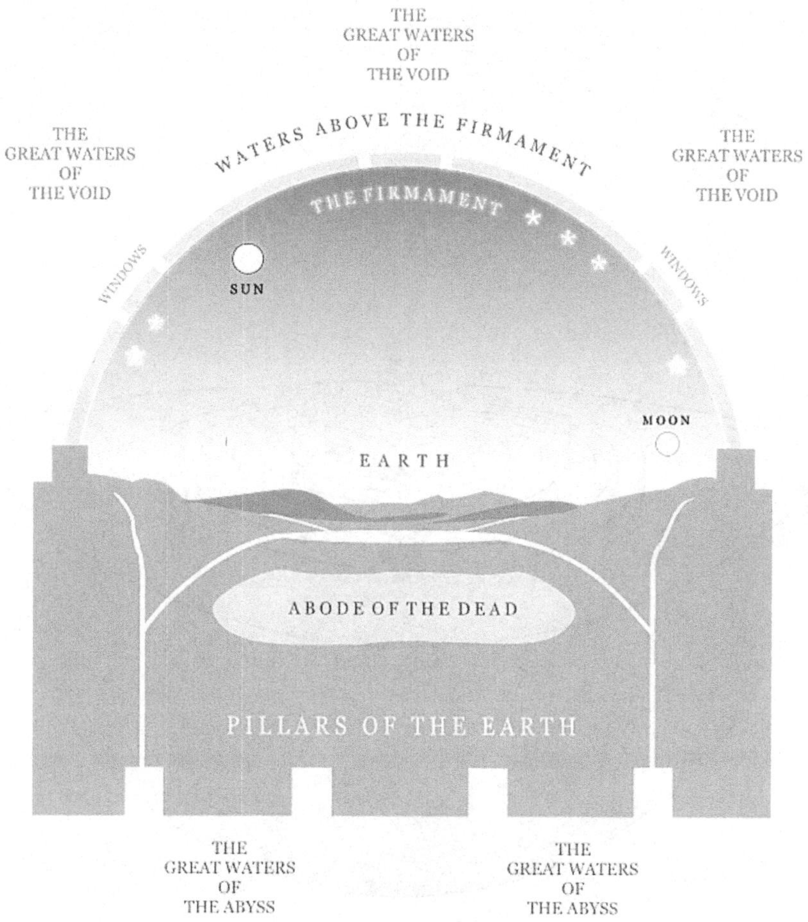

THE HEBREW UNIVERSE

Based on an article by Dr O. C. Whitehouse, as published by Dr James Hastings, in 'The Dictionary of the Bible' (under the word 'Cosmogony') and later discussed in the book: The Earliest Cosmologies, by William F Warren, first published in 1903, pages 19 to 25. The windows in the sky vault served to admit rain, whilst the colour of the sky came by way of the great water of the void which was thought to be limitless in all directions beyond the domain of the earth.

The above illustration
is by the present author

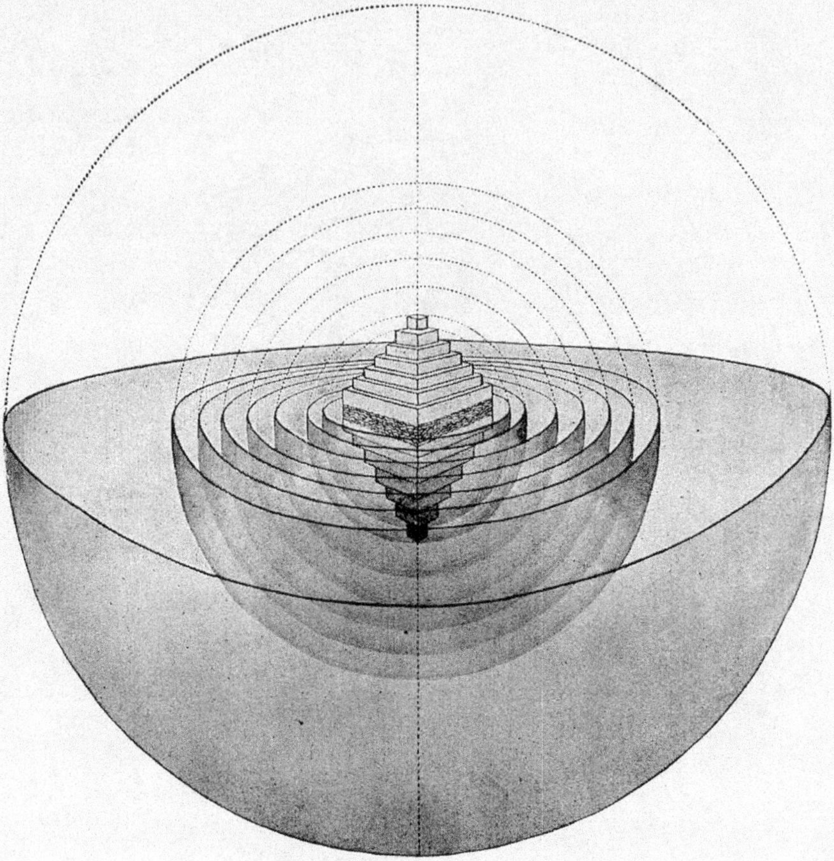

THE BABYLONIAN UNIVERSE

The upright centre line represents the polar axis of the heavens and earth, with north at the top. The two seven-staged pyramids represent the earth; the upper being the 'abode of living men' and the lower being the 'abode of the dead.' The darker central band is the 'waters' which separates the two halves, whilst the seven inner spheres hold the sun, moon and the five known planets. The larger, outer sphere equates to the domain of the fixed stars.

The illustration and the text have been slightly adapted from The Earliest Cosmologies by William F Warren originally published in 1903

side of the sky dome, were all now thought to be a-fixed to a system of slowly rotating crystal spheres. with the earth (as a globe) held free and motionless in the common centre of the system and thus still viewed as the centre of the universe.
Although the number of spheres varied, (Eudoxus of Cnidus: 4th cent BC, used 27 concentric spheres,[8] whilst Callipus: his contemporary used 33,)[9] they all lacked one major factor and that was a sense of the true scale of the universe and its constituents, for at the time the celestial bodies were thought to be relatively small and relatively close at hand.
A perfect example of this is found in a text entitled: The Sand Reckoner, writ by Archimedes in the 3rd cent BC, in which the sphere of the stars is thought to be only 1 light year away,[10] if given in modern day units of measurement.
Fortunately, there were strong minded individuals who were prepared to tackle the shortcomings, Eratosthenes of Cyrene being one such person.
In the 3rd cent BC, using simple geometry, he succeeded in measuring the circumference of the earth.[11] At the time, he was living in the city of Alexandria, where there was a tall column, which he knew lay a certain distance north of his home town of Syrene. He also knew that at noon on a certain day at Syrene, the sun shone straight down into a well. He therefore measured at the appropriate time and day, the length of the shadow cast by the column at Alexandria, and then, by using his knowledge of geometry, he was able to obtain an estimate of the circumference of the earth, out by just a few percent to what we know it to be today viz: 24,900 miles at the equator.
Others who added to the book of cosmological knowledge, were Aristarchus of Samos,[12] and Hipparchus of Nicaea,[12] who lived in the third and second centuries BC, respectively. They each obtained estimates for the distance to the moon using naked eye observations and a little geometry; 79,000 miles,

or (19 earth radii) for Aristarchus, and around 287,000 miles, or (68 earth radii) for Hipparchus - the average distance obtained currently being 238.855 miles, or 60 earth radii.
They, and others, had less success when they tried to measure the distance to the sun, even so, according to Theon of Smyrna, the result obtained by Hipparchus increased the radial size of the known universe to a minimum of 2,550 earth radii, or around 37 times the distance to the moon,[13] which was greater than most of the ancient civilisations ever imagined. As for the distance to the stars: after an earlier suggestion by Aristarchus that the stars were much further away had been rejected,[14] (because with no telescopes yet available, no one could observe an effect known as 'parallax,' - see description below) interest in the distant reaches of the heavens waned, as most observers returned to finding a better solution to explain the orbital motions of the planets, which were still not considered to be fully understood.
Hipparchus took up the challenge and eventually found that if he placed the earth in a central position and re-adjusted some of the planetary motions so that their orbital centre points were slightly off true, the whole system was a better match with what was observed in the night sky, although he also had to introduce what are called *epicycles,*[15] (as illustrated on the next page) to make things work, and in such a way that his concept still allowed room to permit the crystal spheres.

Parallax: The apparent difference in the location of an object when the object is observed from two different positions. A simple example which will demonstrate the effect is to focus on an object with one eye closed and a finger held out in front of your noise. Then, if you alternate which eye is closed, the object will appear to shift position. Astronomers make use of the effect to judge the distance to nearby stars, taking measurements when the earth is on opposite sides of the its orbit.

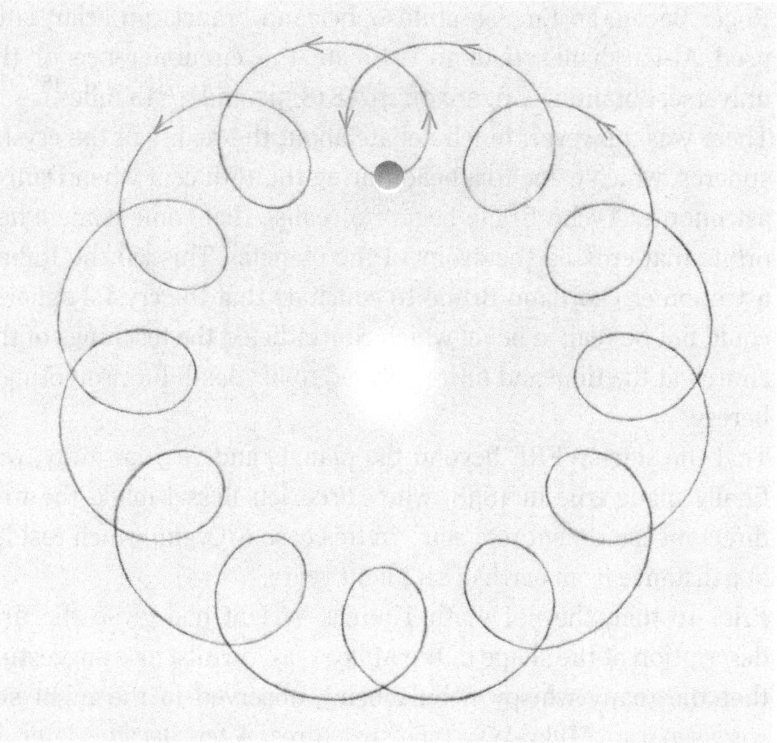

Exaggerated diagram showing a planet sweeping out epicycles in orbit around the sun. Illustration by the author.

His work was later refined by Cladius Ptolemy during the 2nd cent AD, and proved to be so successful that the Ptolemic version held sway over cosmological science for the next 1400 years,[16] until eventually overthrown (in 1543) by the work of Nicolaus Copernicus, a Polish astronomer, who is credited with giving the first correct description of our solar system, with the exception of placing the planets in circular orbits, when, as we now know, they are in fact elliptical.

In the intervening period, improvements were also made to the work of Ptolemy by Islamic astronomers which allowed them to calculate the distance to the sphere of stars, one being Al-Farghani,[17] who obtained a distance of 65,357, 500 miles.

Roger Bacon, an English philosopher and Franciscan Friar, later used Al-Farghani's data to estimate the circumference of the universe, obtaining a figure of: 410,818,517 and 3/7th miles.[18]
There was, however, much debate about the reality of the crystal spheres, which came to a head during the 16th cent when Danish astronomer Tycho Brahe began to realise that some comets had orbits that crossed the orbits of the planets.[19] This led the Italian astronomer Giordano Bruno to conclude that the crystal spheres could not be real;[20] a belief which contradicted the teachings of the church at the time and ultimately led to his death for promoting a heresy.
That the stars WERE beyond the planets and very far away, was finally made true in 1838, when Fredrich Bessel made the first direct measurement to a star,[21] in this case 61Cygni, which resides at a distance from earth of 11.4 light years.
Prior to that, the polymath Thomas Wright had given the first description of the shape of the Milky Way,[22] whilst also suggesting that the many whispy nebula being observed in the night sky were separate Milky-Way type structures. A few decades later, in 1785, William Herschel produced the first map of the Milky Way based on his effort to count the stars,[23] although the ancient Greek Democratus[24] is thought to have been the first to claim the Milky Way to be an assembly of stars, (as suggested in a book by Aristotle called : 'Meteorology,' circa 350 BC.)
Whatever the reality, at the turn of the 20th century, the consensus of opinion amongst most astronomers was that the Milky Way accounted for the entire universe, and it was seen as static. Einstein thought of it as such, as did Sir Arthur Stanley Eddington, the English astronomer, physicist and mathematician who did much to promote Einstein's General theory of Relativity.
Then came the great surprise for in the late 1920's, thanks primarily to the work of the American astronomer Edwin Hubble, we learnt that the observable universe was apparently expanding,

and that our galaxy, the Milky Way, was only one of countless other galactic structures,[25] many of which could now be seen with modern telescopes at ever increasing distances, glowing in the depths of space and possibly out to infinity.

At the time, Eddington's explanation for the apparent expansion was to suggest that the universe had started out as something like a large star cluster[26] which for some reason had destabilized but in such a manner as to instigate the cosmic dispersal. Georges Lemaître, however, (a Belgian priest who was also a professor of mathematics,) proffered instead that the universe had expanded from what he called the Primeval Atom,[27] where all the matter and energy in existence was concentrated into a single volume of space no larger than a pinpoint - before it erupted.

Lemaître called the eruption: the Big Noise, but the British astrophysicist Fred Hoyle (who hated the whole concept) later called it the Big Bang, a term which has been in use ever since to describe the event. In the meanwhile, another person who had long pondered on the origin of the universe and of how the elements might have formed in its earliest moment of creation, was Gerorge Gamow (a Russian émergré.)

He surmised that if the universe had started to expand when it possessed a high temperature and had subsequently cooled over eons of time, then its present-day temperature should be very much smaller but might still be detectable as a remnant of the Big Bang. If so, he suggested, the frequency of its energy[28] would have been lowered as a result of the cosmic expansion, leaving behind a universal abundance of photons.

His prediction, (after some mathematical corrections by his colleagues Ralph Alpher and Robert Herman) later proved to be correct when microwave radiation[29] was detected all across the sky at a temperature of just a few degrees above absolute zero, by Arno Penzias and Robert Wilson, who were working on satellite communication experiments for Bell Laboratories in America.

Their discovery in 1964, is now viewed as strong support for the Big Bang scenario and its consequential expansion of the universe, but for some time before that, there was another paradigm in the running to explain the cosmic expansion, for back in 1948, a radically different theory had been postulated by the afore-named Fred Hoyle, together with Herman Bondi, (the British/Australian cosmologist and mathematician,) and Thomas Gold, (the Austrian born astrophysicist,) which they called : The Steady State model.[30]

In their scenario, which negated the need for a Big Bang event, they held that not only could the universe expand to infinity in all directions, but it could do so whilst its density remained unchanged over time because of a continuous and spontaneous creation of new Hydrogen atoms out of the vacuum of space, which filled in those regions of space left empty by the expansion. (see illustration on the next page.)

The upshot of this meant that the universe had always existed and would continue like-wise forever more with a vista which changed very little over time. However, in the decades which followed the emergence of the Steady State model, astronomical observations began to show that the distribution of certain celestial objects called Quasars and Radio Galaxies, did not match the claims of the Steady State Theory, for the objects could only be found at great distances, rather than at 'all' distances as the Steady State model suggested.

Most cosmologists were henceforth disinclined to give much credence to the Steady State model which by the early 1960's had all but been abandoned,[31] leaving the Big Bang version of events to be subsequently endorsed by the majority of academia - which is still the case today.

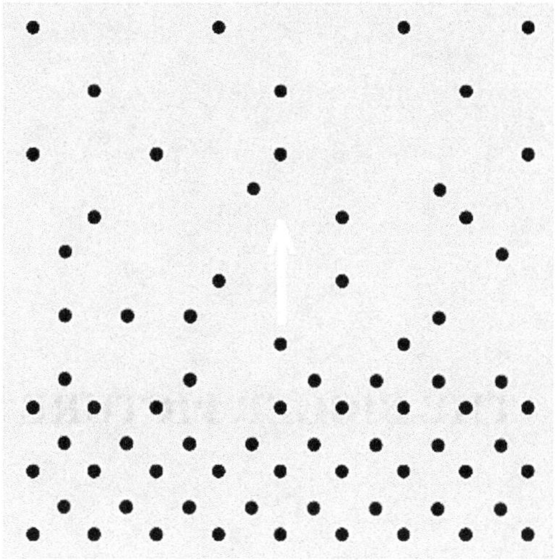

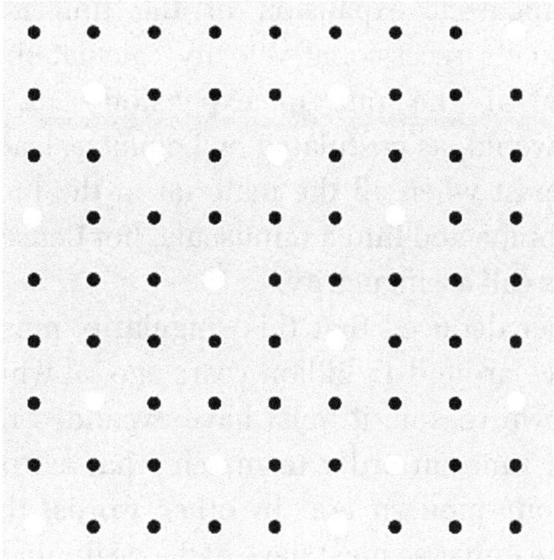

Top: In a simple universe which is undergoing expansion, the number of particles per unit volume decreases with time.

Bottom: In a universe adhering to the Steady State model, new particles, (the white dots,) are spontaneously created to maintain a constant spatial density as the universe expands.

THE BIGGER PICTURE

Once the apparent expansion of the universe had been accepted and its recessional velocity calculated, it was soon realised that if the rate of expansion was theoretically reversed, it would, as postulated by Lemaître, lead to a time in the distant past when all the material in the universe would have been compacted into a minuscule, hot dense state, which cosmologists call a 'singularity.'

It was further deduced that this singularity must have come into existence around 14 billion years ago,[32] at which point, for some unknown reason, it must have expanded rapidly whilst cooling over time, in order to match what astronomers now observe in our modern era. In other words, the expansion suggested the universe must have had a beginning.

Unfortunately for the cosmologists, there are problems with that conclusion, for no one can say exactly HOW or WHY the singularity came into existence in the first place, other than to propose it simply 'popped' into existence 'instantaneously' out of absolute nothingness by 'spontaneous creation,' in what has now become known as the Big Bang event which forms the

basis of the so called 'Standard Model of Cosmology' - the leading theory for explaining the existence of the universe, at least in the scientific community.
Now as you might expect, since the work of Edwin Hubble in the late 1920's, a considerable amount of new observational and theoretical evidence has been put forward to support both the notion of the Big Bang event and the expansion of the universe, with the strongest candidates being : 1) a connection via particle physics which can trace back events to just a few parts of a second AFTER the Big Bang, and : 2) The Cosmic Microwave Background Radiation, or CMBR for short, as detected by Penzias and Wilson back in 1964, which is thought to be the detectable left-over electromagnetic radiation from the Big Bang event.
But is this sufficient to confirm that the universe had a beginning? The short answer is NO! it is not.
Although the above does strongly suggest that there was some kind of major cosmological EVENT in the distant past, it is still a huge leap of faith to say that the Big Bang event was the result of spontaneous creation which brought forth the universe.[33] Despite this, some theorist, have attempted to navigate around this 'leap of faith' by employing new insights of how nature may operate at the smallest scales imaginable, (as in quantum mechanics,) where 'fields of energy' in the vacuum of space itself, are thought to give rise to particles which pop into and out of existence all the time.
They suggest, therefore, that the cause of the Big Bang could have been the result of interactions in these fields of energy creating large quantum 'fluctuations' which eventually arose into the macro-scale universe, (i.e. our observable universe,) where they have never stopped expanding.
Sounds reasonable, you might think. But here's the rub : if you have a beginning, then you also have a before the beginning,

which, to comply with the Standard Model of Cosmology, should be 'absolute nothingness.' It follows, therefore, that if you advocate the existence of energy fields before the Big Bang, then you do not have absolute nothingness, and the universe already exists, all be it not as we see it today.
So, where do we stand at present with respect to the question: Did the universe have a beginning or has it always existed?
Well, let me digress momentarily (before I respond to that question) to clarify what I mean when I refer to both 'the universe,' and to 'absolute nothingness,' as it may help you to better understand the rest of this chapter.

Unless stated otherwise, when I refer to 'the universe,' I mean the 'Whole' universe, viz, that which contains everything, as opposed to the 'Observable' universe, which relates only to what we are able to see or detect with our telescopes and space borne instruments. The 'Whole' universe is thus something which cannot be added too and thereby it contains the potential for everything that can exist, to exist. So whenever you hear talk of parallel worlds, higher dimensions and multi-verses, they are all a part of the 'Whole' universe, and do not exist independently from it.
As for 'absolute nothingness,' I use the following definition:

'Absolute nothingness is LESS than 'empty space' for unlike 'empty space' it has no dimensions.'

What this implies is that there is no such thing as 'absolute nothingness' except as a concept. It is not something you can point to and say "there is an absolute nothingness,' like you can when you want to point out an object, because that which has 'no dimensions' simply does not exist.
Here is a simple example to help further clarity this point.

Imagine an empty box floating freely in space. Although the box is empty, it still has dimensions across the space between its walls, plus the thickness of the walls, and also a dimension in time, (such being the period of time the box exists.)
If, however, the dimensions between the walls of the box, plus the thickness of the walls, plus its dimension in time are all reduced to zero, then the box will no longer have a physical presence in the universe. Hence it will cease to exist.
We can thus further state that with 'absolute nothingness' there are no dimensions in space or time, which thus negates completely any form of reality.
So, returning to the main theme of this book, if 'absolute nothingness' equates to the non-existence of reality, then the universe itself cannot have arisen from 'absolute nothingness' because (as noted above) the non-existence of reality means there is no time and no space - thus, there is NOWHERE for any kind of a creation to take place, and NO TIME when any kind of a creation could occur, and that includes any so called 'spontaneous creation' event.
We are therefore left with only two rational conclusions, viz that, since the universe DOES exist, 1) there must always be a 'somethingness,' as opposed to a 'nothingness,' (which means that the universe has ALWAYS existed without the need for a moment of creation,) and 2) The universe 'overall' is both infinite and static (in the sense of not expanding;) - the former because there is no such thing as 'absolute nothingness,' (so the dimensions of space and time MUST simply go on and on,) and the latter because 'space' is already limitless.
Similar suppositions have, of course, been put forward at numerous time in history. Aristotle, for example, in his book: *The Physics (book 1. p7)* tells us that the world must have always existed because *"everything that comes into existence does so from a substratum. Therefore, if the underlying*

matter of the universe comes into existence, it would do so from a substratum, because it is the nature of matter to BE the substratum from which other things arise. That being so, the underlying matter of the universe could have come into existence only from an already existing matter exactly like itself.

Thus, (by) assuming that the underlying matter of the universe came into existence, (we) assume that an underlying matter already exists. As this is a contradiction, matter must be eternal."

He also argued that a vacuum is impossible because *"material objects can come into existence only in a place. Were something to come from nothing, the place to be occupied by what comes into existence would previously have been occupied by a vacuum, in as much as 'no body' existed. Thus a vacuum is impossible and matter must be eternal."*

In the medieval period, Ibn-Sina (known as : Aviceena, in the west,) a 10/11th Cent AD, Persian polymath, expressed similar beliefs, holding that *"prior to a thing coming into actual existence, its existence must have been possible. Were its existence necessary, the thing would already exist, and were its existence 'impossible' the thing would never exist. The possibility of the thing must therefore, in some sense, have its own existence. 'Possibility' cannot exist in itself, but must reside within a subject. If an already existing matter must proceed everything to come into existence, then clearly nothing, including matter, can come into existence 'ex nilo' as if from absolute nothingness. An absolute beginning for the existence of matter is therefore impossible.**

---

*Note: Both the 'Aristotle' and the 'Avicenna' texts above were drawn from the Wikipedia articles on: 'Eternity of the World' and 'Avicenna' and used as permitted under the 'Creative Commons Attribution-Share/Alike License (*https://creativecommons.org/licenses.by.sa/3.0/)*

In respect of the universe and infinity, Blaise Pascal. a French physicist and mathematician, (1623-1662) offered his thoughts on the matter in his book *'Pansees'* (also known as *'The Thoughts of Blaise Pascal,'* - first believed to have been translated into English by John Walker in 1668.)[34] The relevant passage reads as follows:

"Let man, then, contemplate entire nature in her height and full majesty; let him remove his view from low objects which surround him; let him regard that shining luminary placed as an eternal lamp to give light to the universe; let him consider each as a point, in comparison with the vast circuit by the star (the sun); let him learn with wonder that this vast circuit itself is but a very minute point when compared with that embraced by the stars which roll in the firmament. But if our view stops there, let the imagination pass beyond : it will sooner be wearied with conceiving than nature with supplying food for contemplation.

All this visible world is but an imperceptible point in the ample blossom of nature. No idea approaches it.

In vain we extend our conceptions beyond imaginable spaces; we bring forth atoms, in comparison with the reality of things. It is an infinite sphere whose centre is everywhere and circumference nowhere." *

"But what about the expanding space in our observable universe" I hear you say, "Won't that expansion further increase the size of the larger universe and thereby nullify the notion that the wider universe is essentially static?"

---

* Note : This extract was drawn from *'The Thoughts of Blaise Pascal'* pp 158-159, Chapter II, as can be found on The Internet Archive, University of California Libraries version: cdl americana collection, under the chapter sub-heading : 1. Disproportion of man.
The last line of the passage, are not Pascal's own words, as M Havet, who investigated its origin, found its earlier source to be that of Vinvent de Beauvais, who attributes it to Empedocles, after the monk Helinand, a poet and chronicler of the 12th Cent.

Well, that might seem a possibility until you start to appreciate the difference in scale between the 'observable universe' and the whole universe, for it then becomes clear, (if you accept that there is no such thing as absolute nothingness, and that in consequence the whole universe is 'limitless') that any part of the universe which may be expanding (such as our observable universe) will be so insignificant in size compared to the whole universe, that the region of expansion could never cross the whole expanse of the wider universe, regardless of how much time was involved.

The most likely scenario, therefore, is that any expanding region of space will simply disperse its excess energy into the larger universe, and will, thereby, eventually cease to expand.

In any event, the notion of an eternal and infinite universe can here be seen to have advantageous consequences for the Standard Model of Cosmology, since it provides an energy source for the Big Bang event and its aftermath, (assuming such really took place,) as well as explaining what the observable universe is expanding into, (since it cannot be expanding into 'absolute nothingness' - which, as already stated in this book, has no dimensions)

But should we really accept the above on the strength of the scenario outlined, or can we add weight to the concept of an eternal and infinite universe still further?

Well, there are in fact a few possibilities that we should certainly consider, and these I will now endeavour to explain in the following chapter.

THE RED SHIFT & AGE QUANDARY

According to mainstream cosmological thinking in our present era, the further we look into the far reaches of the universe (and in consequence, the further back in time since the speed of light is finite,) we should see galaxies which have less mass and are younger in age, than those galaxies which are closer to us. The younger galaxies should also be moving away from us with a much higher 'red-shift,' (i.e. speed of recession) than those galaxies which are generally older and closer to us, since such has been deemed to be the natural order of things ever since the work of Edwin Hubble demonstrated that the observable universe is apparently expanding.

Until recently, such was taken almost as a given, since the consensus of opinion was that big galaxies acquired most of their mass via mergers with smaller galaxies,[35] which all takes time, hence their greater age and lesser red-shift, equates to a slower speed of recession.

However, the results of a study (published in 2014) of 16 massive and well evolved galaxies with high red-shifts has called that belief into question, a state of affairs made more perplexing by the equally startling revelation that these same galaxies, with an average mass of 50 billion times the mass of the sun, appear to have an average estimated age of around 800 million years,[35] with the bulk of their mass seemingly accreted just a few hundred million years after the Big Bang.

If such galaxies were to be found in large numbers, it could well overturn our current scenarios for galactic evolution, particularly when the early universe (in the Big Bang model) is generally seen as an environment mainly composed of hydrogen and helium with only low concentrations of a few other types of elements.

Whilst all the above are certainly puzzling for those who support the standard model, there are two unorthodox possibilities that come to mind to explain the above galactic anomalies, viz, 1) the universe is much older than mainstream science currently thinks, (since such would account for the time needed for the anomalous galaxies to fully form,) and 2) some (or all) of the anomalous galaxies are actually part of a wider universe that extends beyond our observable universe, with the anomalous galaxies having previously existed in that region of space where the Big Bang subsequently took place and where-after the expansion of the space of our observable universe sent the anomalous galaxies flying away from us at some era in the distant past, thus accounting for their currently observed high red-shift motion.

Another study[36] (undertaken by a team of astronomers from Nanjing University of China) of two galaxies much closer to our home base, also proved surprising in that the galaxies were found to have an elemental make-up much akin to the type of galaxies as would normally be expected in the early history of our observable universe, when, as noted above, the types of element in existence were of limited variety.

It thus suggested, therefore, that because the two nearby galaxies involved in the study were not as well supplied with

as wide a variety of elements as other nearby galaxies, star formation in the early history of our observable universe may not have been as efficient as previously thought, meaning the early galaxies (in general) may have taken longer to mature than our current understanding of how galaxies have evolved, although it has to be said, there could be other details presently undiscovered in the very early galaxies, which may yet come into play to alter the above view once the power of our telescopes has been further improved.
In the meanwhile, the cosmological red shift, one of the long held assertions of The Standard Model of Cosmology, is being challenged by the revival of a radically different cosmological model known as 'The Tired Light theory,' (to be discussed in the next chapter) which, if proven to be correct, would have far reaching consequences for how we view the universe since it would negate the need for the cosmological red shift.

The cosmological red shift is simply the term cosmologists use to describe the 'apparent' expansion of space based on a shift in the position of lines which appear in the spectrum of the light from most galaxies.
Where such galaxies appear to be receding from us, (as a consequence of space expanding) the lines in the spectrum will appear to be shifted from their normal 'rest' position towards the red end of the spectrum. And the greater the speed of expansion, the greater the shift of the spectral lines.
Although the light from some galaxies (and stars) are said to be 'blue' shifted,' where the lines in the light of their spectrum will be shifted to the blue end of the spectrum, this is because such galaxies (and stars) are themselves moving towards us via a process which cosmologists call a 'Doppler Shift' (which can be confusing to some since Doppler shifts can also be redshifted.)
Cosmological red shifts on the other hand, are always thought

by cosmologists as signifying the expansion of space away from an observer, even though it is not, in fact, direct evidence that space is expanding in any way. It is just an idea applied to the movement of light in the manner as suggested by the way sound waves on Earth alter in wavelength when the sound of an object such as the whistle of a train changes in pitch as the train moves towards or away from us.

Even Edwin Hubble, who set the ball rolling with respect to a possible relationship between the expansion of space and red shift - as a consequence of his discoveries back in the late 1920's - was hesitant to endorse such a notion as can be seen from the comments he wrote in his book 'Realm of the Nebulae,' (first published in 1936) which runs a follows:

"The crucial data are surrounded by uncertainties. By pressing the data to the limit of their tolerance, always in one direction, we might force the velocity-shift into the framework of the surveys. The universe would then be small and filled with matter to the threshold of perception.

On the other hand, if the interpretation of the velocity-shifts is abandoned, we find in the velocity-shift a hitherto unrecognised principle whose implications are unknown.

The expanding universe of general relativity would still persist in theory, but the rate of expansion would not be indicated by the observations."

The truth is redshifts can come about in other ways which have nothing to do with the expansion of space, for eg, via 'Gravitational Redshifts' (see diagram at the end of this chapter,) as when photons of light in leaving the sun on a journey to the Earth move from a strong gravitational field to a weaker one - losing a little of their energy in the process which in consequence lowers the frequency of the light - or, as noted previously, via a process as to be outlined in the 'New Tired Light' theory in the next chapter.)

It is also perhaps worth noting here - for the sake of clarity, that the physical objects of the universe, the stars, planets, galaxies and atoms, (which are themselves bound together by the various forces of nature, such as gravity, the electrostatic force, and the strong and weak nuclear forces, are not effected by the expansion of space, which all adds to the strangeness of the universe.

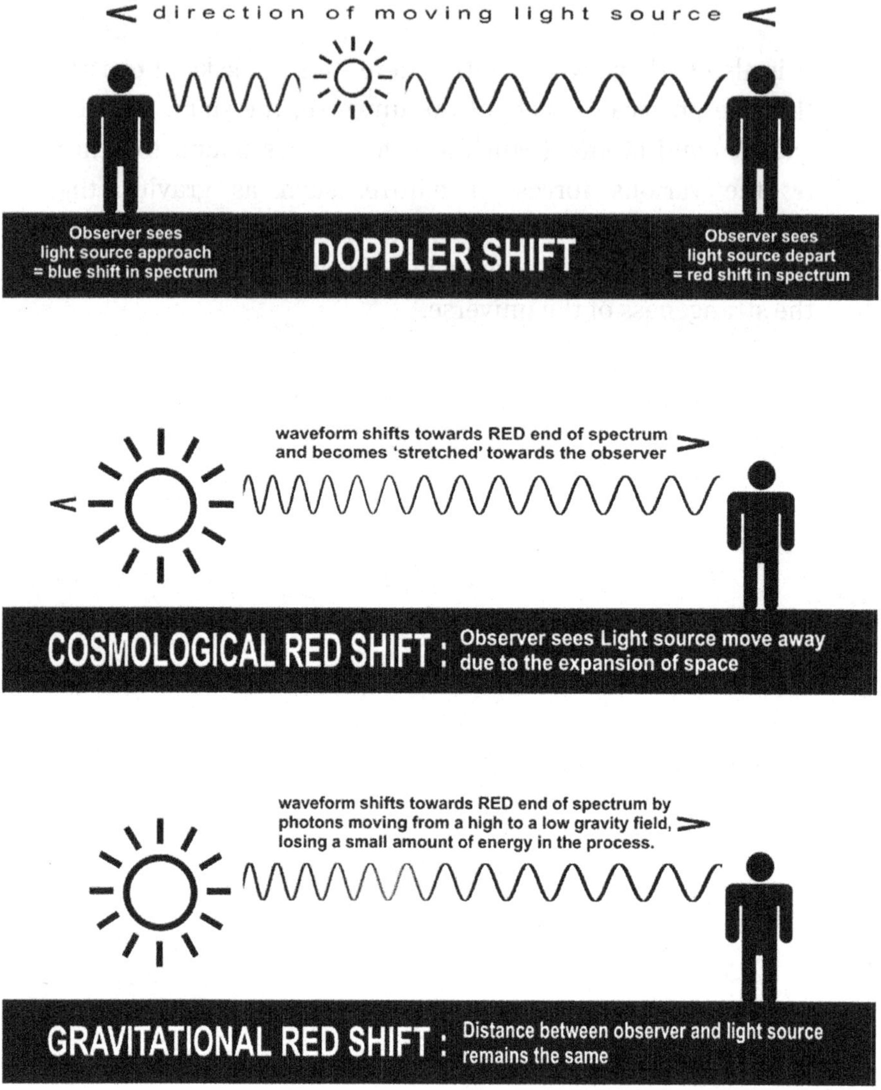

The Doppler Shift, also known as the Doppler Effect, is named after Christian Doppler, an Austrian physicist and mathematician, who, in 1842, proposed in his published work 'Concerning the Coloured Light of Stars' that the frequency of a wave is dependent upon the relative speed of the source and the observer. Six years later, the French physicist Armand Hippolyte Louis Fizeau predicted the red shifting of electromagnetic waves.

source : Wikipedia : on Doppler Shift, & Fizeau. *Author's diagram*

THE 'NEW' TIRED LIGHT THEORY

In simple terms, the 'Tired Light' theory may be defined as an alternative explanation for the red shift 'distance relationship,' which, in the science of cosmology, one associates with the expansion of space.

It was first proposed by the Swiss/American astronomer Fritz Zwicky, in his 1929 paper entitled: On the Red Shift of Spectral Lines through Interstellar Space,[37] which posits that photons gradually loose energy as they move through space via collisions with other entities in the spatial environment, thereby producing an effect that reveals itself as a reddening of distant objects relative to the those closer to an observer.

After due consideration, however, Zwicky abandoned the idea having come to appreciate that any scattering of light would result in blurred images of distant objects. He suggested instead that :

"one might expect a shift of spectral lines due to the difference of the static gravitational potential at different distances from the centre of the galaxy. This effect, of course, has no relation to the distance of the observed galaxy from our own system, and cannot therefore, offer any explanation of the phenomenon discussed in this paper"

But if Zwicky had misgivings about the validity of his own Tired Light theory, research into the concept remained on-going elsewhere and indeed it has undergone a re-evaluation in more recent times, thanks to the input of new insights into particle physics and other areas of modern science.

Today, the new Tired Light theory is seen in terms of a static universe where:

"photons of light from distant galaxies are absorbed and re-emitted by electrons in the plasma of intergalactic space and where, with each interaction, the electron recoils." [38]

What this amounts to is that each photon loses energy if it encounters an electron, resulting in the photon's frequency being lowered, its wavelength being increased and its spectrum signature being shunted towards the red end of the spectrum. And since the number of photon encounters with electrons will increase as the photons make their way across the vastness of space, with more encounters occuring at the red end of the spectrum than at the blue, the photons are said to have undergone a 'red-shift.

If such red-shifts are then discerned with different values, that information can be used as a kind of distance indicator since if the photons of light from a galaxy twice as far away undergoes twice as many collisions, it should have a higher red-shift than the photons of light from a nearer galaxy, once other factors are taken into account such as the value of 'collision cross-sections,' (see below,) and the estimated number of electrons per cubic meter of intergalactic space.

"Collision cross sections, which have units of area, represent a probability that a collision between two particles will take place. In a photon-electron collision, there are only two possible outcomes. Either the photon is absorbed and not re-emitted, or the photon is absorbed and a new photon is emitted."

source: An Explanation of Redshift in a Static Universe: Lyndon Ashmore: vol 6 Number 2, page 4 : Proceedings of the National Philosophy Alliance. 2010

The New Tired Light theory can therefore estimate a value for the Hubble constant[38] (in terms of its red-shift to distance relationship) and this has proven to be consistent with the value as quoted in the Standard Model of Cosmology.
The plasma of space (as referred to earlier) relates to the vast clouds of hot gas where some or all of its constituents are separated into electrons and ions which are free to move about independently. It is the energy which these same electrons acquire and then emit as secondary radiation after their interaction with the travelling photons which, (in the Tired Light theory,) is said to account for the pattern of the Cosmic Microwave Background Radiation, where the more clumpy areas are thought to equate to the plasma clouds themselves.[39]
Although the Tired Light theory remains an unorthodox concept, it did attain a degree of credibility when it predicted the temperature for the Cosmic Microwave Background Radiation of 2.8° Kelvin when other models which favour the Big Bang scenario were contemplating diverse temperatures of between 5°Kelvin and 50°Kelvin.[40]

The problem for supporters of the Tired Light theory is that in order to compete with the widely accepted Standard Model of Cosmology, there are a number of issues which the Tired Light theory needs to resolve, not least its miss-match with the Standard Model when it comes to the surface brightness of galaxies. Here's how things stand with regards to that :
In a Tired Light universe, the surface brightness of galaxies are said to remain essentially constant, relative to the 'area' of sky which they encompass. Thus, if two similar galaxies are situated at different distances, although the more remote galaxy will look fainter (as observed from Earth,) it will still have the same 'brightness to area' relationship as the brighter galaxy. This is not the case, however, with an expanding universe, for although the surface brightness of galaxies also

diminishes with distance, the area of the sky which the galaxy appears to cover is somewhat larger than it really is because when the light from the galaxy begins its journey to Earth, the first emitted photons are closer to us than the receding body of the galaxy which creates a false impression resulting in a different 'brightness to area' relationship.[41]
The Tired Light theory also needs to explain the alleged signs of 'time dilation' within the so-called 'light curves' of remote Type Ia supernovae, as such may be indicative of 'expansion.'

Type Ia supernova represent the final evolutionary stage of white dwarf stars which suddenly rise to a brightness equal to over a billion suns exploding all at once over a short time scale. In the process, all the material of the star is scattered out into the universe, whilst the brightness subsides much more slowly over the coming number of weeks, to finally reach a stage that makes it difficult to be seen. This is actually the fate of the vast majority of all stars.[42]
When the data from observing such events is compiled into the supernova's light-curve diagram, (basically a graph with a time axis running horizontally and a brightness axis running vertically) researchers can study the information the graph contains to see if there is any extension in the amount of time the brightness of the supernova takes to fade, compared to the graph of a similar supernova at a different red-shift. If a difference is found, it is seen as signifying time dilation due to universal expansion.
The idea was first proposed by the astronomer Olin Chaddock Wilson in August 1939, in his paper: Possible applications of Supernovae to the study of the Nebular Red Shifts.[43]
It is a technique which in recent times has been replaced by a new and more selective approach which focuses on the time intervals when various elements make their appearance in different spectra - to off-set worries about Malmquist bias[44]: a

natural tendency for observers to favour (cosmologically speaking) brighter objects over those which look dimmer, because the latter are more difficult to see, and also because it is now known that not all Type Ia supernova attain the same level of brightness.[45]

From all the evidence thus far gained from such research as discussed above, it appears to show that in the past, our observable universe was indeed expanding - judged on the basis of the supernovae light curves which have high red shifts. On the other hand, whether time dilation is also present in 'low redshift' Type Ia supernovae (which occur closer to Earth) is still open to question as results achieved thus far are somewhat contradictory.[45]

Oddly enough, research dealing with the light curves of Quasars[46] has also raised an enigma, for to echo the words of Michael Hawkings, of the Institute for Astronomy, University of Edinburgh:

"... after an analysis of over 800 Quasar light curves to measure time scales of variation at different redshifts, (for eg: for Quasars at 6 billion and 10 billion light years away,) the expected effects of time dilation were found to be absent, the energy distribution at high and low redshifts being essentially identical.....

If it is accepted that time dilation is not observed in Quasar light curves, then some departure from conventional cosmology is necessary to explain it ...... whilst time dilation in supernovae light curves has yet to be convincingly established."[47]

With respect to the latter, British physicist Lyndon Ashmore, suggests that any supernova time dilation (if real) might be down to dispersion,[48] given that white light is a composite of multiple colours. Perhaps it undergoes 'pulse broadening' as occurs with light when in transit along an fibre optic cable as a result of the light's individual colours moving at different velocities.

Whatever the truth, the issues with supernovae and Quasar stellar light curves highlights that there is no conclusive proof in favour of or against the conjectured universal expansion, at least at the present time.
If it should turn out that there really is no expansion of space in the observable universe, then it would clearly undermine the current Standard Model of Cosmology, and add strength to the notions of both the Tired Light/Static Universe and the Eternal/Infinite Universe as proposed by this book under the banner of : The Non-Zero Universe, (which can actually work with or without any expansion - simply on its core assertion that there is no such thing as an 'absolute nothingness'.)

THE BASELINE OF REALITY

Let us now take a different tack by asking the following question: What are the minimum requirements for something to be called: a universe?

Well, obviously we need 'space' since this provides us with 'dimensions,' and if we have 'space' then we also have 'time' since the two are inseparable. We can also then claim to have the framework for a 'reality' since even if 'space' is entirely devoid of all matter and energy and nothing ever happens, time could still be regarded as passing, even if its 'duration' could not be measured - due to the lack of any discernible 'change.'

But since we know the universe exists and IS constantly in a state of change, we must assume there has to be something more significant with what we call 'empty space,' to bring about the various 'changes' that we know are continuously happening in nature - and the only thing we know about that fills that need is 'energy.'

We can therefore make the case that the whole universe must

have always had some form of energy, rather than just empty 'space,' for the processes of nature to have evolved in the manner we think they have.

This energy may be the 'quantum-foam' that physicists speak about, which is said to permeate the cosmos in a constant state of motion at the smallest scales of nature, or it may be the 'dark energy' which cosmologist believe is the root cause for the expansion of space in our observable universe. It may even be something totally different that we haven't even detected yet, after all, to quote Sir Martin Rees, the current Astronomer Royal, "Absence of evidence does not mean evidence for absence."

Whatever the truth, it would appear to be the case that at the very core of reality, 'reality' itself is simply pure energy, not as something which has to be created IN space, but rather as something we currently describe AS space, which perhaps interacts with our brains to provide us all with self awareness as well as our picture of the universe.

The difference in the above wording may be very subtle, but it does have a value since it would seem to put an end to the age old question of 'what came before the beginning' of the universe. It also opens up the possibility that the Big Bang which is said to have triggered the expansion of our observable universe, may simply be a common feature of the whole universe if considered over very long time scales, creating bubble-like structures of energy which expand and disperse, to constantly change not only the surface appearance of the whole universe, but also its very nature at different scales of reality.

So, is there anything which could undermine the notion of an eternal universe besides the supposed spontaneous creation from nothingness concept.

Well, as history often tells us, any theory or supposition which does not align with expert opinion or is seen as contrary to a majority held view - is inevitably fated to be met with opposition and criticism, and the eternal universe theory is a prime example of this, since it has long been frowned upon by both religious institutions and by the academic elite; the church because they see God as the creator of all things, and the academic elite because they believe there are two major 'factors' at play which negate an eternal universe, these factors being : 1) 'Entropy,' and 2) 'The Principal of Causality.'

Taking Entropy first. This basically deals with the dispersal of energy into its surroundings whenever permitted to do so. A simple example would be a hot cup of tea cooling down over time. The tea will continue to cool until it has the same temperature as its surrounding environment, whereat it would reach a state of 'equilibrium,' whence no further transfer of heat could occur unless heat was re-supplied from an external source. There are, however, currently, three ways of describing 'the entropy of a 'thermodynamic system,' to give it its more scientific title, and these 'systems' have been classified as : 'Open,' 'Closed,' or 'Isolated.'

An 'Open' system can exchange both energy and matter with its surroundings, whilst a 'Closed' system can only exchange energy with its surroundings. The third option, an 'Isolated' system cannot exchange anything with its surroundings, so the whole universe can be viewed as an 'Isolated' system, since nothing surrounds the whole universe where an exchange of any sort is possible. The problem here, however, is that the whole universe could also be classed as an 'Open' system, since it is also 'not isolated from anything,' given that by definition, it contains 'all things in creation.'

Because, as this last point shows, there is no definitive way of expressing what the notion of entropy is for all circumstances, or "*for a system out of equilibrium, even when in a stationary*

state,"[49] some scientists believe it is *"rather presumptuous to speak of the entropy of the universe ..... and its major constituents when such have never been in equilibrium in their entire existence."*[50]
Others, of those who object to an eternal universe, believe the universe will evolve along the lines as described at the start of this section, as per the cooling cup of tea, so that eventually equilibrium will be reached everywhere, whence all motion will cease and the universe will effectively die.
The truth of the matter is, however, no one can say for certain what the outcome will be, but if the universe IS infinite in size (as it has to be, if there is no such thing as absolute nothingness,) then equilibrium will never be attained across the entire universe, because such would require an infinite amount of time, and that equates to 'time without end.'

Let us turn then to the other 'factor' mentioned earlier, viz : The Principle of Causality, sometimes referred too simply as : First Cause.
This argument against an eternal universe asserts that : *'Every event has a cause and thus a beginning.'*
Whilst such appears to hold true in everyday life, when we seek to discover the cause of the whole universe, we find ourselves eventually at an impasse, since if there was a beginning to the whole universe, (as maintained by 'The Standard Model of Cosmology,') then we would expect there to be a 'before the beginning.' Even the Greek philosopher Paramenides of Elea, came to that very same conclusion way back in the 6th century BC, when he held that : *"Nothing comes from nothing."*[51]
So, is there a way to resolve this issue?
Well, you could turn to religion and put the whole of creation down to the hand of God, (ignoring who created God,) or, if like the present author, you require a more secular answer,

then perhaps one way to consider the matter would be to turn to the simple use of numbers, since it is often said that mathematics underpins all of science, so :

Lets say ‘0’ equals : Non- Existence
and ‘1’ equals : The Universe.

We can deduce from the above that:

since ‘0’ does not exist, it cannot give rise to ‘1’
and ‘1’ can never emerge from ‘0’
therefore, ‘1’ must already exist
since the universe exists

We may also deduce that infinite regression is impossible
since before ‘1’ there is only ‘0’
which equals Non-Existence
so ‘1’ must always exist since the universe exists
and ‘1’ can never emerge from ‘0’

Despite the above adhering to common sense, there are still some who reject the suppositions preferring the notion of a universe which owes its very existence to the posits of quantum physics, (as mentioned earlier in this book) where, at the smallest scales in nature, sub-atomic, energetic particles are said to pop into and out of existence via ‘continuously fluctuating energy fields’ said to permeate the universe in the vacuum of space. Some scientists take this conjecture and use it to proclaim the universe has arisen from ‘nothing,’ pointing to the Heisenberg uncertainty principle - a principle of nuclear physics - to endorse their belief.
In fact, what the ‘uncertainty principal’ actually tells us is that we can never know ‘accurately’ what the values of two related

variables are at the same time, such as, for example, the amount of energy a quantum particle retains with respect to a certain value of time. In other words, the more certain we are of the value for one of the variables, the less certain we are of the value of the other variable.

Heisenberg's equation for the 'uncertainty principle' does however inform us that if the time value is very short, the energy value must be very high - and vice versa, and it is this aspect of the 'uncertainty principal' that some people point too, to claim the universe came from 'nothing,' for the following reasons :

As hinted at previously, in quantum physics, empty space is no longer considered as a void of 'nothingness,' but is, instead, visualised as a collection of 'energy fields,' each of which 'fluctuates' across a range of values considered to be their lowest energy state, which it maintains without ever attaining complete rest. This leaves them with a trait known as 'Quantum Harmonic Oscillation.' In short they jitter.

This is what makes it difficult to assess their energy value with respect to the time the energy exists (as noted above.)

Fortunately, the energy values of REAL particles can be represented by a simply diagram showing energy values increasing in steps from the bottom upwards, (see diagram on the next page,) with the first step up being labelled: Ground State, which equates to the lowest energy state for all REAL particles which exist in the universe.

Energy still exists below the 'ground state' level, but generally not at a sufficient magnitude to be assigned the status of a REAL particle. It is therefore, considered as the background energy which is said to permeate throughout the universe, although it still has disturbances of its own which seem to vary in value completely randomly.

If, however, these disturbances become excessive, their value can rise above the ground state value of real particles which

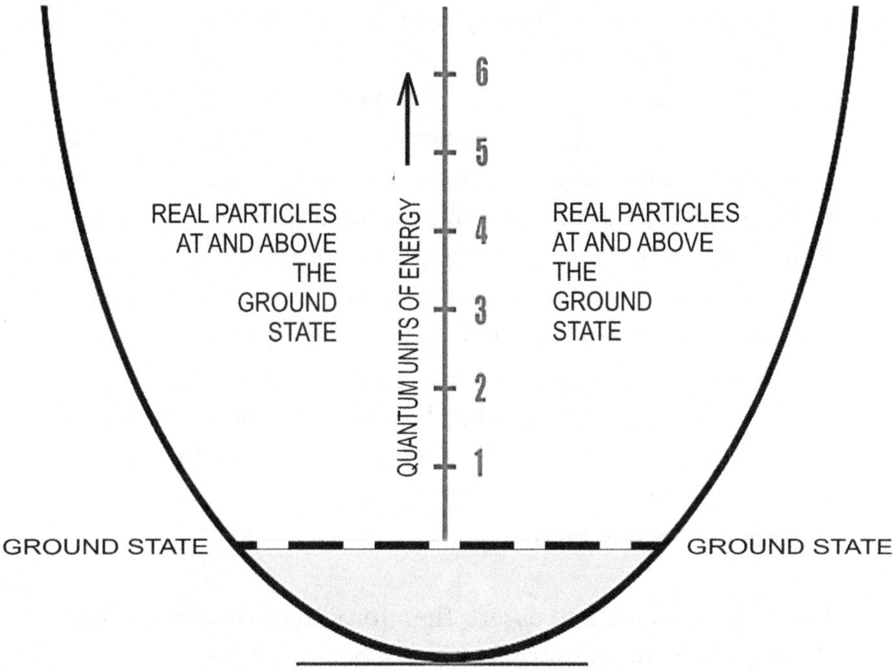

The shaded area below the Ground State contains no real particles. Instead, it is full of low level fields of vacuum energy which is in constant turbulence. If the value of the vacuum energy increases beyond that of the ground state energy level, new pairs of particles, (one positive and one negative) are formed out of the energy fields, whereafter they annihilate each other almost at once.

According to Longman's 'Physics Handbook' page 148 : A Quantum is "the smallest amount of matter or energy that there can be."

turns the energy from the disturbances into pairs of tiny new particles, (one positively charged and the other negatively charged,) which then almost immediately annihilate each other. This sequence tells us three things:
1) If you put enough energy into one place, the energy will convert into 'mass.'
2) The new particles are NOT popping into and out of existence from a 'nothingness,' they are being converted into mass, from the energy which already exists as the background energy that fills the universe. (Thus the universe already exists.)
And 3) Since the new particles may also possess attractive or repulsive forces, it is possible that at some point in the distant past, some of the new particle survived long enough to group together to create a large enough energy source that started to expand, thereby creating what has become known as the Big Bang, (but negating the 'singularity' problem in the process) whilst continuing to evolve over billions of years, into the universe as we see it today.

It seems reasonable to assert, therefore, that, contrary to what the detractors of an Eternal universe believe, it is this notion of a permanent background energy in the so called 'vacuum' of space, which actually supports the supposition of an eternal universe as noted above as well as earlier in this book.
We may thus conclude, at least in theory, that the universe had NO beginning and will continue to exist forever more, evolving at its own pace and in its own mysterious manner which we have yet to fully understand and appreciate.

THE LINGERING PUZZLE

Although academia now has a plethora of cosmology models besides the widely accepted Standard Model of Cosmology, all have only been possible with the inclusion of assumptions or speculations which are either not provable at present, or cannot be counted upon with a high degree of certainty.

There are two main reasons for this. The first, (according to the Standard Model,) is that we are limited in how far we can peer into the universe due to the expansion of space across our observable universe, the outer extent of which recedes from us with a velocity which increases with distance. Because of this, any galaxy that is 5 billion light years away from us, will have a recessional velocity that equals the speed of light, whilst hypothetically it is positioned on the edge of a spherical boundary, which scientists call 'The Hubble Sphere,' with the Earth at its centre. Beyond that, space will be receding from us at faster than the speed of light,[52] which effectively blocks our scrutiny of any region of space which may exist beyond, at least for a time, since if 'space' continues to expand, so too will the size of the Hubble Sphere, thereby increasing the size of the universe we will be able to observe.[53]

Currently, the observable universe is thought to have a radius of around 46 billion light years, which includes an allowance for the apparent expansion of space. That estimate may, however, be somewhat erroneous due to the possibility that some of our stellar and galactic distance measurements are being incorrectly assessed - at least in theory. Here's how it all stacks up :

Suppose we have a star with a life span of 10 billion years, situated at a distance of 20 billion light years from Earth. We can say that that star will radiate light in all directions - such that the light will extend outwards from the surface of the star to a maximum distance of 10 billion light years - before the star ceases to shine.

The starlight will then continue its journey out across the cosmos for as long as it is able to do so, with one thin beam of its light (10 billion light years long) heading in our direction.

In the meanwhile, because the (now dead) parent star is 20 billion light years from Earth, a gap will develop between the surface of the dead star and the tail-end of the light-beam, with the gap continuing to grow as time goes by.

Now suppose that when that beam of starlight finally reaches the Earth, it is observed by an astronomer who decides to measure the distance to the star. What would he or she discover?

The answer is probably quite a lot about the age, colour, surface temperature and the chemistry of the star with the aid of a spectroscope, since all such information is contained within the spectrum of the starlight. As for the distance to the star, that might not be so straight forward for here we are faced with a problem.

Since the astronomer can only have knowledge of the star and its position in the cosmos via its light, any calculation made is potentially an incorrect one for any circumstance similar to that presented above, since the astronomer cannot know how far away the dead star really is - because of the gap which developed behind the tail-end of the starlight. We can thus express the problem like so:

If the life-span of a star
is shorter than the time needed for its light to reach the Earth
the stellar or galactic distance measurement
may very likely be in error.

In other words, there exists the possible potential that some of our estimated distances to the remote stars and galaxies may actually only be distance measurements to their celestial light sources, rather than a true measurement to the actual celestial bodies.

If true, such would have obvious implications for the currently accepted age of the universe for even if just a few examples were the remnant light from stars or galaxies which are no longer shinning, it would mean the source of the light would have travelled a great deal further than we thought, making the universe larger and older than we thought, which in turn would undermine the Standard Model of Cosmology, leaving open the possibility that the whole universe really could be infinite in both time and size.

Now, some reader may think, with respect to the latter comment, that if the universe as a whole is eternal, such that even the light from no longer active stars and galaxies is still whizzing around the universe, then why is the sky not ablaze with a constant light?

There are several reason to explain why such would not be so, the most obvious one being that for any given period of time there are only a finite number of stars and galaxies emitting light, non of which exist forever.

Yes, much of the light from the non-longer active stars and galaxies would still be extant, but given the infinite age and size of an eternal universe, much of the ancient light would have been absorbed by the

‘matter, in the universe, and much of it will have already passed through the volume of space which the Earth and the solar system currently occupies, to the extent that we can no longer detect it.
It should also be remembered that the further away the emitted light travels from it’s parent source the more dispersed the density of the photons will become per unit volume of space, making such light harder to detect by eye, plus their energy value will be reduced to a lower frequency - perhaps to beyond the range of visible light - every time the photons pass through a strong gravitational field.
It can be seen, therefore, taking all the above points into account, that most likely only a small percentage of all the photons ever emitted will eventually reach us - the observers on Earth.
Unfortunately, we will probably never be able to know for certain if the universe is truly infinite in time and space since it is not possible to cross an infinite span of space. The best we can hope for would be for the expansion of space (if such is real) to slow down to a velocity below the speed of light. That at least might allow us to known if there is indeed more space beyond the limit of our ‘observable universe’ for under such a condition light from any galaxy that might currently exist beyond our observable universe would be able to enter our observable region, even though if would take billions of years for such distant light to make the journey to Earth, before any astronomer still in existence would be able to see it.

The other main reason for the uncertainty with respect to the correctness of our cosmological models is that the two great theories of the modern science, namely General Relativity (which deals with the world as we experience it at the macroscopic level,) and Quantum Physics (which studies the world of the very small,) cannot be brought together into one all encompassing ‘theory of everything,’ which suggests that there are either gaps within our knowledge of how the universe operates, (almost certain to be true however much we learn,) or we have been barking up the wrong tree for goodness knows how long, which has led modern cosmology into its present day crisis.

As ever, only time has the answers to every mystery and whilst this book and the speculative content therein makes no claim to hold a greater wisdom than any other, speculations per say, are not necessarily a bad thing, since they can help to prevent academic stagnation, for if such were NOT the case, then the larder of cosmological thought would be pretty bare today, given that there are probably more speculative notions in the annals of modern cosmology and particle physics than in most other fields of modern intellectual endeavour.

In any event, the search for new knowledge continues apace and the whole universe beckons.

NOTES & REFERENCES

1) The Early Hebrew Conception of the Universe: an article as published 1909, in: The White & Blue (the student newspaper of Brigham Young University): Ralph Vary Chamberlin: Vol XIII, number 11, pp 84-88.
In the above article, Professor Chamberlin expresses the view that the Hebrews regarded water as "the foundation of all things," and that, "before the earth was formed," they supposed that "water alone existed in space."

2) The Earliest Cosmologies: 1903: William F Warren: pp 19-28

3) A History of Zoroastrianism: Mary Boyce: Vol 1, p132.
"The earliest accounts which were pre-iron age, described the sky as: an empty shell, perfectly round, made of stone, passing beneath the earth as well as passing over it."

4) Coming of Age in the Milky Way: Timothy Ferris: 1989: p 20.

Note: In the Jewish Encyclopaedia, the word 'Firmament' is said to be a translation of 'raqia,' meaning, in biblical Hebrew, to 'beat out,' or 'spread thinly,' whilst 'heaven' is a translation of 'shamayim.'
The Hebrew regarded the earth as a 'plain' or 'a hill' figured like a hemisphere, swimming on water. Over this is arched a solid vault of heaven to which are fastened the 'lights,' viz 'the stars.' So slight is its elevation, the birds may rise to it and fly along its expanse.
(source:Cosmogony: http://www.jewishencyclopaedia.com/view.jsp?jewishencyclopaedia.com artid=807&letter=C#2736.
retrieved: 3 February 2020.

5) The Earliest Cosmologies: William F Warren: 1903: pp 34-37.

6) Mesopotamian Civilisation on : ancient civilisations.com

7) Measuring the Universe: Albert van Helden: 1986: p 4.

8) Coming of Age in the Milky Way: Timothy Ferris: 1989: p 27.

9) Aristotelian Explorations: G E R Lloyd: 1996: p 173.

10) Coming of Age in the Milky Way: Timothy Ferris: 1989: p 38.

11) noted in article: Windows to the Universe: as modified by: Randy Russell: 31 July 2007.
source: Window2universe.org: retrieved: 5 February 2020.

12) noted in article: Ancient Greek Astronomy: Denis Erkal: giving distance to the moon by Aristartchus, and Hipparchus.
source: people.ast.cam.ac.uk - University of Cambridge: retrieved: 5 February 2020.

13) noted in the article: Hipparchus: Lunar and Solar Theory, distance, Parallax, Size of the Moon and Sun.
source: www.primidi.com/hipparchus: retrieved: 6 February 2020.

14) noted in the Wikipedia article: Aristartchus of Samos.
retrieved: 5 February 2020.

15) The Eagle Book of the Universe: 1960: p 14.

16) The Eagle Book of the Universe: 1960: p 14.

17) Measuring the Universe: Albert van Helden: 1986: pp29-31.

18) Planets, Stars and Orbs - The Medieval Cosmos: Edward Grant. 1994: p 433.

19) The Eagle Book of the Universe: 1960: p 17.

20) The Eagle Book of the Universe: 1960: p 17.

21) noted in an internet article about: Fredrich Wilhelm Bessel: Britannica Encyclopaedia.

22) An original Theory or a new Hypothesis of the Universe: 1750: Thomas Wright: p 62.

23) noted in the wikipedia article on William Herschel: section: 2.8: Further Discoveries.

24) noted in: Democratus: an article by Mark J Joshua: 15 Februay 2011: Ancient History Encyclopaedia.
retrieved: 4 February 2020.

25) Coming of Age in the Milky Way: Timothy Ferris: 1989: p208.

26) Coming of Age in the Milky Way: Timothy Ferris: 1989: p211.

27) The Eagle Book of the Universe. 1960. p156.

28) Coming of Age in the Milky Way: Timothy Ferris: 1989: p213.

29) Coming of Age in the Milky Way: Timothy Ferris: 1989: p214.

30) Wikipedia: Steady State model.

31) Wikipedia : Steady State model.

Note 1) *additional* : According to an article by David Castelvecchi, entitled: *Einstein's lost theory uncovered. pub in 'NATURE' 24 Feb 2014. Volume 506, pp 418-419.* 'Albert Einstein had also considered a steady state theory as a possibility for explaining the expansion of the universe in a 1931 manuscript, but soon abandoned the idea.'

Note 2) *additional* : According to Wikipedia, in its essay on the Steady State model (of cosmology,) 'During the 13th Cent, the philosopher Sigar of Brabant authored the thesis : The Eternity of the World, which postulated that there was *'no first man or specimen of any particular kind,' and thus no beginning to the universe which was thus eternal.*
Needless to say his views did not sit well with the church and were condemned by the Pope in 1277.

32) At the time of this writing there is no consensus for the expansion rate of the observable universe and in consequence of that, to the time of the Big Bang.
According to Dr John Huchra of the Harvard-Smithsonian Centre for Astrophysics, there have been over 500 values published for the rate of expansion up to the year 2008. as noted in a New York Times article by Dennis Overbye (as posted on line 18 of August 2008, under the heading : The Struggle to measure the Cosmic Expansion,) - and the uncertainty continues.
In 2013, (for eg,) a team of European scientists studying data relating to the so called relic radiation of the Cosmic Microwave Background, as gained by the European Planck Space Telescope, concluded the rate of expansion to be 67Km/sec/megaparsec, which places the age of the universe at 13.8 Billion years. However, when a team from the Space Science Institute, (located in Boulder Colorado,) later attempted to verify the result, they attained instead a higher value of 74Km/sec/megaparsec, which

another team almost equalled with a value for rate of expansion of 73.3Km/sec/megaparsec, placing the age of the universe at 12.5 Billion years.

More recently, a team from the Max Planck Institute for Physics, (Munich) using a different measuring technique, arrived at a new higher rate of expansion of 82.4 Km/sec/megaparsec, which lowers the age of the known universe to 11.4 Billion years, (although the true age value may be somewhat higher, as the measuring technique employed (the only option at the time) is know to be less reliable than other methods. None the less, there is a growing belief amongst the astrophysical community, that the discrepancy in the results is increasingly suggesting that the computational models which are employed when analysing the data may be where the fault lays, rather than the observing techniques, and studies are thus under-way to improve the theoretical models.

Data sourced from : an Associated Press article 12 Sept 2019, as posted on line by NBC News, headlined : 'The Universe may be Billions of years younger than we thought.'
also from : an article on the website :INSIDEHOOK, written by Joe Dziemianowicz : 18 May 2019, 'The Universe may have just gotten a billion years younger.'

NOTE: *V*esto Slipher is credited with collecting the first observational evidence that the universe is in-fact expanding.
source : Coming of Age in the Milky Way : Timothy Ferris : 1998 : p206

33) With respect to the Big Bang as the genesis for the creation of the universe, Neil Turok, a theoretical physicist stated :
"People have inferred that time began then, but there really wasn't any reason for the inference."
Noted in an article by James Randerson : 5 May 2006, in The Guardian : headlined: One Big Bang, or were there many?
source :https://www.theguardian.com/science/2006/may/5/spaceexploration.universe. (retrieved : 17 Feb 2020)

34) Classical Probability in the Enlightenment: Lorraine Datson: p34, (1988)

35) 'A Study of Massive and Evolved Galaxies at high Redshift': by H Nayyeri et al: AstroPhysics Journal: p794: 68 (10 Oct 2014.)

36) 'Galaxies in the Early Universe were less than Stellar Star Makers': an article by Nola Taylor Redd: October 2014: via Space.com

37) 'On the Redshift of Spectral Lines through Interstellar Space': research paper by Fritz Zwicky: Proceedings of the Natural Academy of Science: Vol 15, pp773-779 (1929)

38) 'An Explanation of Redshift in a Static Universe': a research paper by Lyndon Ashmore: Proceedings of the Natural Philosophy Alliance: Long Beach 2010, p6.

39) 'An Explanation of Redshift in a Static Universe': a research paper by Lyndon Ashmore : Proceedings of the Natural Philosophy Alliance: Long Beach 2010, p1.

40) 'Reflections and Thoughts on Tired Light': published essay by: M Moore & J Dunning-Davies: p2 : arxiv.org/pdf/0707.3351.

41) Noted in Wikipedia article on: Tired Light: History & Reception section: The Tolman Surface Brightness Test.

42) Info drawn from You Tube lectures by Michel van Biezen: Astronomy/Supernova: lecture 3, and Astronomy/Exceeding the Limit: Lecture 10.

43) Astrophysics Journal: 90: pp634-636 (1939)

44) First described in 1922 by Swedish astronomer Gunnar Malquist. *source : Wikipedia.*

45) 'Supernovae Ia Light Curves show a Static Universe': Proceedings of the Natural Philosophy Alliance: a research paper by Lyndon Ashmore: vol 9, p18 (2012.)

46) Quasars (also known as Quasar Stellar Objects) are believed to be exceedingly bright objects which "inhabit the centre of galaxies.' They were first detected in the 1950's as sources of radio wave emissions, and are found over a broad range of distances, with some almost 30 billion light years away. *source : Wikipedia.*

47) 'On Time Dilation in Quasar Light Curves.': Monthly Notices of the Royal Astronomical Society: a research paper by M. S. R Hawkins: printed: 13 April 2010. (additional source: Article by Lisa Zyga: physics.org.com: 9 April 2010.

48) 'An Explanation of Redshift in a Static Universe.': a research paper by Lyndon Ashmore : Proceedings of the Natural Philosophy Alliance: vol 6, No2, p6 (Long Beach 2010.)

49) Giovanni Gallavotti: 1999: Statistical Mechanics-A Short Treatise. p 290.

50) Walter Grandy Jr: 2008: Entropy and the Time Evolution of Macroscopic Systems: p 151.

51) Although he lived in the 6th Cent BC, Parmenidies dictum that "Nothing comes from nothing" is thought to have first appeared as a text in Aritotles book: The Physics. (4th Cent BC.) *source: Wikipedia.*

52) This does not contravene Einstein's assertion that 'nothing can move faster than the speed of light,' because his assertion

is only applicable to things moving THROUGH space. Space itself is free to move at any speed it likes, as far as we know.

Note : The expansion of space is cosmologically referred to as 'The Hubble Constant' and is often written in scientific texts as: H^{o}.
It is a measure of velocity quoted in Kilometres per second for every Megaparsec of distance, (where 1 Megaparsec equals 3.26 million light years,) and 1 light year (as the name suggests) is the distance travelled by light in 1 year at a speed of 186,000 miles per second - when moving through a vacuum.)

53) For this to be true, you need to picture a galaxy just beyond the boundary of the Hubble Sphere.
In such a position, the galaxy would be moving away from us at FASTER than the speed of light by the expansion of space in its region. The LIGHT from the galaxy meanwhile, would be making its way towards us (at the speed of light) whilst at the same time being dragged away from us by the faster than light speed expansion of space in its neighbourhood, (so that overall, the galactic light would still be moving away from us.)
It follows, therefore, that if the galactic light coming towards us is ever to become known to us, the Hubble Sphere will need to expand faster than the actual speed of that 'incoming' galactic light, at which time the Hubble Sphere will be able to overtake the incoming galactic light, allowing the latter to proceed Earthwards more directly. It would of course still be billions of years before we saw such light, due to the apparent expansion of space in our region plus the distance involved.
In the meanwhile, we can continue to gaze in wonder at all we see in the night sky and be grateful for the view.

FURTHER READING

P. Coles : The Routledge Companion to the New Cosmology : *(Taylor & Francis)*

J. Gribbin : Companion to the Cosmos : *(Orion Books)*

M. S. Longair : Our Evolving Universe : (Cambridge University Press)

J. D. Barrow : Theories of Everything : *(Oxford University Press)*

F. J. Dyson : Infinite in All Directions : *(Penguin)*

E. R. Harrison : Masks of the Universe : *(Macmillan)*

C. Sagan : Cosmos : *(MacDonald)*

J. Silk : A Short History of the Universe : *(W. H. Freeman)*

S. Weinberg : The First Three Minutes : *(Basic Books)*

A. Guth : The Inflationary Universe : *(Jonathan Cape)*

M. Rees : Before the Beginning : *(Touchstone)*

J. North : History of Astronomy and Cosmology : *(Fontana)*

F. Hoyle : Frontiers of Astronomy : *(Heinemann)*

M. Davidson : From Atoms to Stars : *(Hutchinson)*

H. Shapley : Galaxies : *(Harvard University Press)*

M. Davidson : Astronomy for Everyone : *(Dent)*

D. S. Evans : Frontiers of Astronomy : *(Sigma Books)*

M. Chown : The Afterglow of Creation : *(Faber & Faber)*

I. Morison : Introduction to Astronomy & Cosmology : *(Wiley)*

Neil deGrasse Tyson : Astrophysics for People in a Hurry : *(W W Norton)*

B. Clegg : The Quantum Age : *(Icon Books)*

S. Clark : The Unknown Universe : *(W W Norton)*

B. Cox and J. Forshaw : A Journey through the Cosmos :
(Penguin)

B. Clegg : Dark Matter & Dark Energy : *(Icon Books)*

F. Close : The New Cosmic Onion : *(CRC Press)*

S. Carroll : Quantum Worlds & the emergence of Spacetime :
(One World Publications)

J. al-Khalili : Quantum - A Guide for the Perplexed : *(Phoenix)*

V. A. Rubakov & D. S. Gorbunov :
Introduction to the theory of the Early Universe
(World Scientific)

W. M. O'Neil : Early Astronomy from Babylonia to Copernicus
(University of Chicago Press)

A. Liddle & J. Loveday : Oxford Companion to Cosmology
(Oxford University Press)

A. Ananthaswamy : The Edge of Physics : *(Duckworth Overlook)*

D. W. Sciama : Modern Cosmology : *(Cambridge University Press)*

E. Seife : Alpha & Omega : *(Transworld Limited)*

J. Gribbin & M Rees : The Stuff of the Universe : *(Cornerstone)*

E. W. Kolb et al : Inner Space - Outer Space
(University Chicago Press)

R. Waterfield : A Hundred Years of Astronomy
(Macmillan)

O. Struve & Velta Zebergs : Astronomy of the 20th Century
(Macmillan)

N. Calder : Violent Universe : *(Viking)*

B. J & Priscilla Bok : The Milky Way *(Wiley)*

P. Doig : A Concise History of Astronomy : *(Phiosophical Library)*

C. A. Whitney : The Discovery of our Galaxy : *(Knopf)*

J. S. Trefil : The Moment of Creation : *(Scriber's)*

S. I. Shklovskii : Stars, Their Birth, Life, and Death
(Freeman)

D. W. Sciama : Modern Cosmology : *(Cambridge University Press)*

J. Narlikar : The Structure of the Universe
(Oxford University Press)

F. Hoyle : The Nature of the Universe : *(Harper)*

P. W. Hodge : Galaxies : *(Harvard University Press)*

G. Gamow : The Creation of the Universe : *(Mentor)*

J. Silk : The Creation & Evolution of the Universe
(Freeman)

S. Mitton : Exploring the Galaxies : *(Scribner's)*

D. MacLagan : Creation Myths : *(Thames & Hudson)*

L. Kuhn : The Milky Way
The Structure & Development of our Star System
(Wiley)